Radiant Energy: A Guide to Wireless Power

© May 2017 Michael C Ellis

This document may not be reproduced in any form or in any language, written or spoken, digital or print except for parody, constructive criticism, review or educational use as defined by law.

The author makes no claims, medical or otherwise. Radiant Energy is not a substitute for proper medical treatment. Any use of this information is at the readers own risk. The reader agrees to release the author and the publisher from any liability by reading the following document and performing any experimentation based on the information provided. All of the information provided in this document is open source, provided without warranty or guarantee. The safety of wireless energy has not been medically evaluated and the reader accepts any risk of life or limb. The author provides advice to the best of his abilities. Do not conduct these experiments if you or a loved one have a pacemaker or medical implants. Keep all electronic devices away from children and adults. Dangerous voltages may exist or form on metal objects, coils, capacitors, or batteries. Harmful radiation may be created. Supervise children operating any electronic equipment.

Absolute Resistance Press

Monona, Wisconsin

Dedicated to Mr. Nikola Tesla, and all of his wonderful inventions we take for granted, and to those in need across the globe.

Radiant Energy

Preface

"I am Ironman!" I always wanted to say that. That movie, the first one, inspired me to become an electrical engineer. Tony Stark was awesome. He was smart, funny, had all the girls, and all the technology. He also had free energy. So it began. However, I was interested in energy long before then. I was 13 when I noticed the planet needed energy. Just that. Energy. More of it. Free energy. Maybe I was 16, I'm not really sure. I had a crush on Amy Lee of Evanescence. I had amorous plans to sweep her off of her feet with my free energy technology and marry her someday, as I drove home from Walgreens at 10PM blasting her music in my Talbot yellow Fiero on winding back country roads. Free energy was real to me, though. It was all around us in nature. I was sick a lot as a kid. I stayed home with sinus infections and tonsillitis until I had surgeries to correct those problems. I can still remember watching those alien and UFO shows on daytime TV. They obviously ran on free energy, the UFOs. The visitors from space seemed to have the technology, why didn't we? Well we did. 100 years ago. It was lost to history. Nikola Tesla invented it. Of course I didn't know that then. Ignorance is not bliss, but it gives you purpose. I began studying everything I could get my hands on to solve this energy crisis. This was in 1999. Orgone seemed to be a match. Everything about it made sense and it worked, it was real. It wasn't until recently that I learned that Nikola Tesla discovered that the Earth was enveloped by terrestrial magnetic waves at approximately 12Hz (Meyers). What happens when a magnetic wave, basically an EMP, hits conductors nested between insulators (orgone accumulators)? It accumulates electric charge. Mystery solved. I continued to build these mysterious orgone devices for what they were worth, fully trusting that they worked however unaware of their mechanism of physical action for years before finally learning that they worked on some type of magnetic plasma, mostly through remote viewing. Then came the age of Facebook and the internets. A series of tubes. I slowly discovered that I was actually not Michael Ellis as I began heading off into the world of college studies. During my second psychiatric hospitalization I read Gerry Vassilatos' book Secrets of Cold War Technologies, after haphazardly grabbing it off a shelf for something to read while I was in the hospital. During some down time in the hospital I decided to remote view the 1) Optimal Radiant Energy Circuit and 2) Tesla's Ether to Electricity Device. I wanted to test how my medication was affecting my psychic abilities. I was surprised to discover that the radiant energy circuit had a battery ground, and that the ether to electricity device resembled a Joe Cell. I later remote viewed Tesla's Pierce Arrow circuit and this became the basis of the wireless receiver. The rest is in this book.

I would like to thank Frederick JP Langheim, MD for all of his help in working with me through my struggles, Jan Schubert MSCW, also very helpful, Brianna Krieg and the rest of the Yaharans, my mom Tammy Ellis and dad Christopher Ellis, brother Nick Ellis PhD and sister in-law Alisha Ellis for putting up with my mind-altering experiments in the apartment and basement of the house, Tom and Sarah for being so accommodating, Contessa Miller, Don and Carol Croft, Sensei Dennis, Alberto Rodriguez, PhD, Terrance O'Laughlin, PhD, and Melanie Herzog, PhD.

- Michael C Ellis

Radiant Energy

Introduction

This is it. This is the answer to the energy crisis. I have spent years trying to channel this out of the ethers and now I have finally done it. The following pages hold Nikola Tesla's technology that you haven't seen before. Wireless power in all of its glory. This comes with great responsibility. On the footsteps of September 11th, when Nikola Tesla's weaponized technology was used to bring down the World Trade Center towers in horrifying display, when the towers were reduced to vapor and ash, metal cut into neat sections, and vehicles burnt to a crisp, all while papers and passports were literally unaffected, firefighters trapped beneath dust, and victims made mockery of literally stripping their clothes off to avoid the anti-personnel effects in the gaping hole of the towers. It is just too much to deny the horrifying abuse of this technology at the hands of the elite. The answer, of course, is to put it squarely in the hands of the people. That is why I have recovered this technology. We need this free energy technology to advance our civilization. There lies a silver lining under the bloodshed this technology has been burdened with. Limitless power from the stars, heating and lighting from the ethers. Life in the most extreme and remote climates. Space exploration. First world life for all. Just some of the benefits of this technology. We are just scratching the surface. Imagine an all-electric automobile fleet, completely powered by the ethers (read stars), reducing the carbon footprint of the goods that travel to your doorstep every day. It is possible to live well in harmony with nature. It is possible to reduce suffering across the globe, accommodating those in need, and to light the world, as Tesla envisioned. Our world is in needless war for resources we could be harvesting from space. What lies beyond our solar system? Is it possible to build intergalactic civilizations run on infinite energy from the stars? Free energy is here, are you ready?

Radiant Energy

Getting off the Grid

The first part of your Radiant Energy package is the Wireless Power Receiver (WiPoR). This is a generator designed to perform basic functions such as heating, lighting, cooling, running core appliances, pumping, heating and purifying water, and charging batteries and mobile devices. This package is going to be accessible for 24/7 use, reliable in all weather conditions, and useful for conventional utility outages. The ultimate purpose of this utility is to replace conventional utilities with wireless etheric utilities.

The WiPoR is powered by longitudinal waves in the ethers that impact a conventional radio circuit with slight modifications to channel etheric energy also called orgone, prana, or chi. This ether circuit generates a Radiant Energy impulse signal that is used for various applications. While this may look like a standard electrical circuit with conventional electrical properties, note that this does not use conventional electricity. Even though electrical components are used, they serve a different function than conventional electrical engineering. All electricity generated is ancillary to the actual function of the circuits. We are engineering ether, not electrical engineering. You will soon see this in action as you follow this guide. The most difficult thing to accept whilst engineering this technology is that you will not be able to measure this energy with standard equipment (aside from byproduct electricity generated). You will, however, be able to FEEL this energy with your body. You are powered by the ether, just as much as these circuits are.

True Wireless Power

Wait, no wires? Yes, no wires. The ether is everywhere, electromagnetic, superluminal (faster than the speed of light), and gaseous. In fact, the energy you are about to use is sourced from the Sun as it impacts the Earth, and is an inexhaustible solar electromagnetic particle influx. It penetrates the walls of your home, your body, and the very Earth itself. The planet is a vast reservoir of energy as it flows through the ground. The reason we use a battery is because the chemicals act as a sink for this energy. In fact, a battery is all that is needed to run this circuit, no ground connection is required. This would allow for mobility in a vehicle such as an aircraft, train, boat, or car. Possibly even your mobile devices. By raising a conductive plate, ball, or pipe, you enable a connection to a wireless power mesh network, like Tesla had dreamed of. This is also your input, where you download power from the ethers. Radiant Energy is just that, radiant. It radiates. With the right setup, not just hard wired, it is possible to transmit energy wirelessly. You can create radiant force fields of etheric energy. It is possible to build up reservoirs of energy in the medium with the right circuit. No battery storage is required, you can store energy in the ether just as Tesla planned to do with Wardenclyffe!

Just a word of safety before I reveal the circuit. Keep your circuit tuned to 1MHz or more (Vassilatos). Otherwise heat or pain may be broadcast instead of useful ether. That means using the equation $f = \frac{1}{2\pi\sqrt{LC}}$. Where electricity is generated you run the risk of injury. Please use safe isolation of electrical force fields from the body to avoid injury or death. Forewarned.

Radiant Energy Circuit

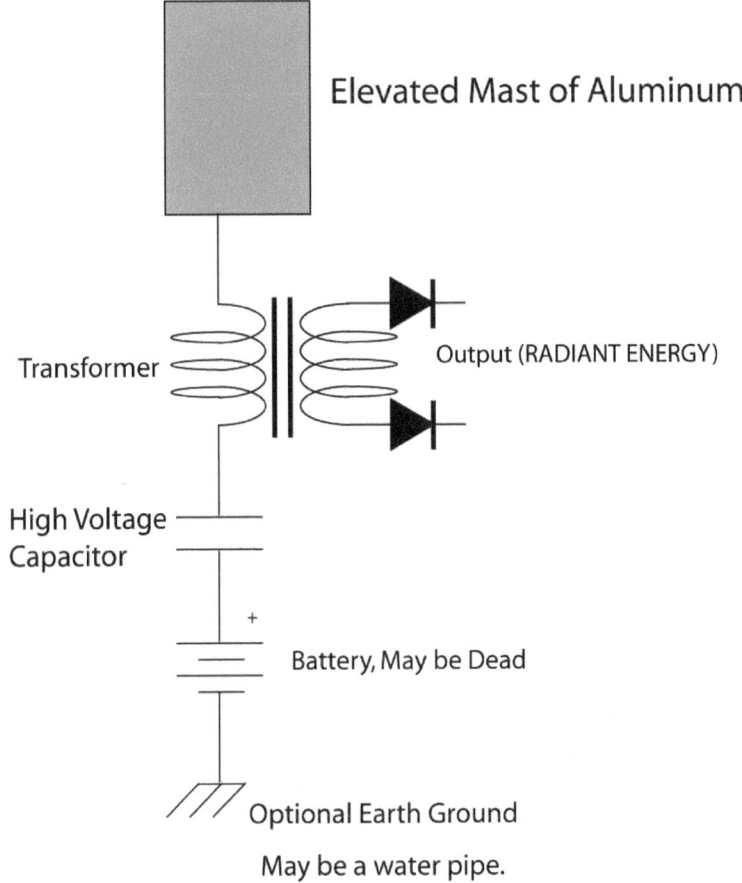

Now What?

Now your first question is probably, "what do I do with this circuit?" A simple wirewound resistor attached to the negative of your battery will guarantee you an unlimited supply of alternating current. It is really that simple. An inductor, wirewound resistor, or transformer of any type across the battery will provide useful voltage. This is the circuit Nikola Tesla's car used to run a 90mph AC motor (minus the diodes). Multiple transformers may be used to introduce multiple phases or simply more utilities/frequencies, grounded to the same battery. Different frequencies can be used to provide: pain (0Hz-400Hz), anti-personnel heat (200Hz-400Hz), warmth (400Hz-800Hz), attention (800Hz-200MHz), awe (200MHz-800MHz), cooling (400MHz-800MHz), light (800MHz-900MHz), or intuition (800MHz and up), if one so desires. A series of diodes will intensify the radiation and radiate it out into the environment as they resonate at their resonant frequency. The heating, which is infrared or solar-like heat can be used to warm tissues from within. In fact, the warmth produced by radiant energy might kill viruses such as the common cold, pneumonia, or even the flu by assisting the body's immune system. Radiant energy may even have undiscovered antitumor effects at certain frequencies according to psychic intelligence. Artificial penetrating light can be simulated by using higher frequencies or by reversing the battery, projecting from the aluminum mast. This may also affects the weather, which seems to respond to concentrations of electrical energy in the atmosphere. In that regard, note that if you accidentally create a negative field of energy, a good rain will neutralize the field. The author learned this the hard way.

Radiant Energy

On the Topic of Electricity

Do not, and this is a warning, try to rectify the output into a battery. This attempt was met with a minor boom or implosion in the ether when the author attempted this, with two diodes facing opposite directions. It literally shook the house and startled the author's parents. One must understand that extreme forces are at work with this circuit. Do not take this lightly simply because this looks like a radio circuit. One will note the force field effects when working with this circuit. The effects will grow and ebb and flow to fruition. This may be measured with a simple digital multimeter on the AC setting. While this appears to be a weak electrical circuit, the consequences can be devastating if misused. Note that this puts great responsibility in the hands of the reader. Act wisely or not at all. Danger exists if the circuit is shorted back onto the mast, as this produces electron radiation. This may charge batteries if properly rectified, but produces a field effect of electron radiation that may cause bodily harm. This is a dangerous practice and is to be avoided at all costs and is quite painful (imagine getting electrocuted from within). These electric dartlets are what endangered Nikola Tesla's life when he was struck over the heart and nearly died (Vassilatos). Even though your mast is not likely generating electric fields, it is best practice to keep your mast and other equipment out of reach of children and adults alike. Do not, however, be afraid of the etheric force fields. They are quite safe to add and draw power from without harming biological functioning. Nikola Tesla proved this by engulfing himself in a glow of plasma of Radiant Energy during one historic demonstration.

DO NOT DO THIS

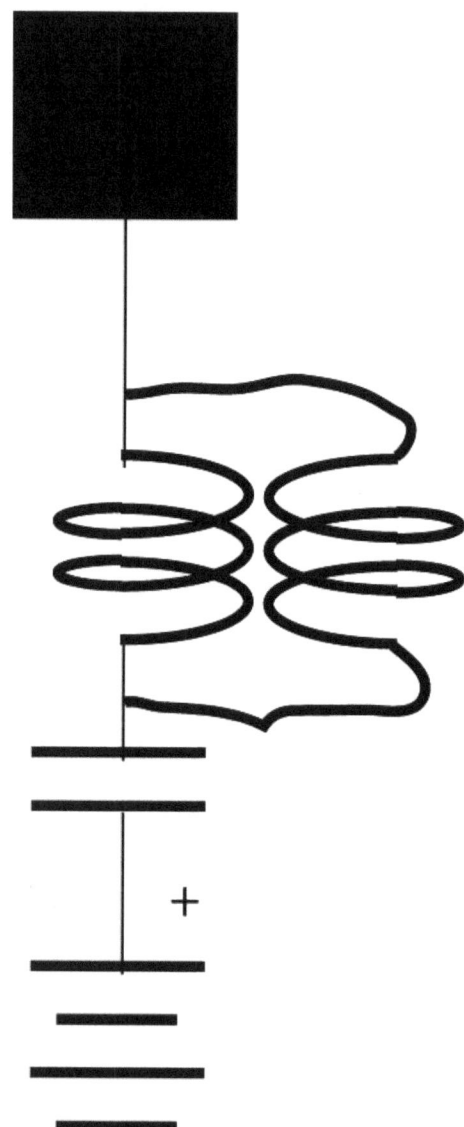

Radiant Energy

Using Your Power

All that is required to draw power from a force field is an inductance and a good connection to a ground or battery negative, in this case. Charging batteries and mobile devices is simply an extension of the alternating current. Once an alternating current is established in a receiving circuit, a direct current may be obtained. The circuit to do this is a simple rectification of the waveform and smoothing by capacitors and inductors. A potentiometer is used to adjust the voltage to the correct value for charging. An automated charging controller may be used if desired.

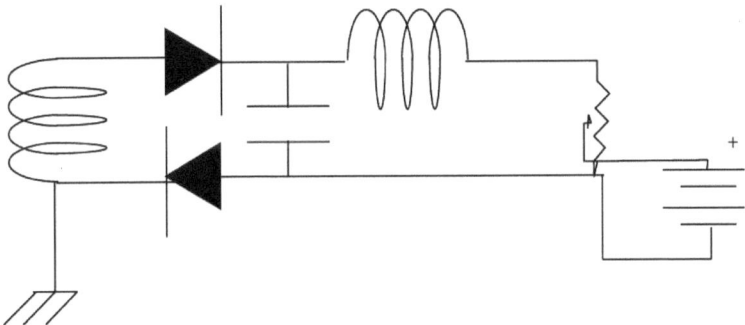

Step up or step down may be accomplished by use of a series of wirewound resistances and coils. This would enable one to access various levels of AC voltages for various appliances. Now this could be run directly from the force field or it could be run off of an inverter from a battery bank. However we will use the force field as an example since this minimizes infrastructure usage. Note that Tesla himself noted power gains whilst stepping voltages from one magnitude to another.

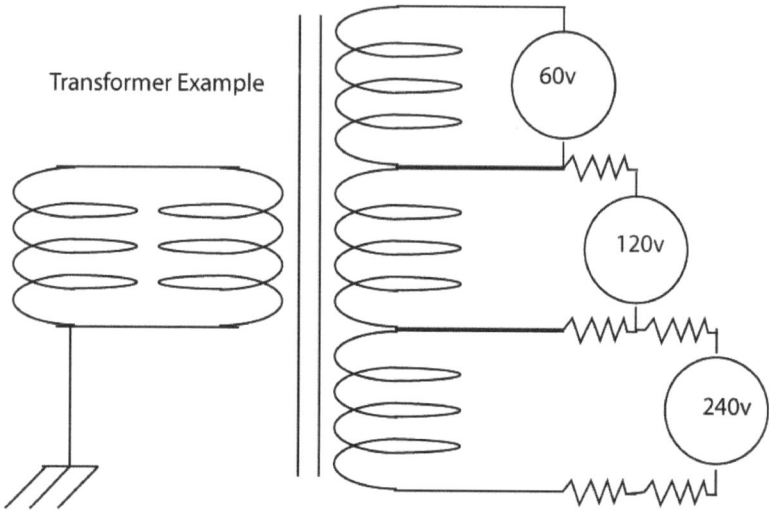

The Ether Valve

The Ether Valve (EVe) is the next-generation of Tesla technology. This was one of Nikola Tesla's late inventions that died with him. It never actually reached public use. The author managed to get ahold of this much like the previous invention via remote viewing. This greatly resembles the Joe Cell. Ether is channeled through an ion valve consisting of nothing more than nested iron pipes and transformer to charge a battery or capacitor. The power cell is then used to run a device. Tesla used this power anything from toys to top secret government spacecraft. It is quite powerful and you deserve to have a piece of the pie! Here is the circuit diagram.

Radiant Energy

ETHER VALVE

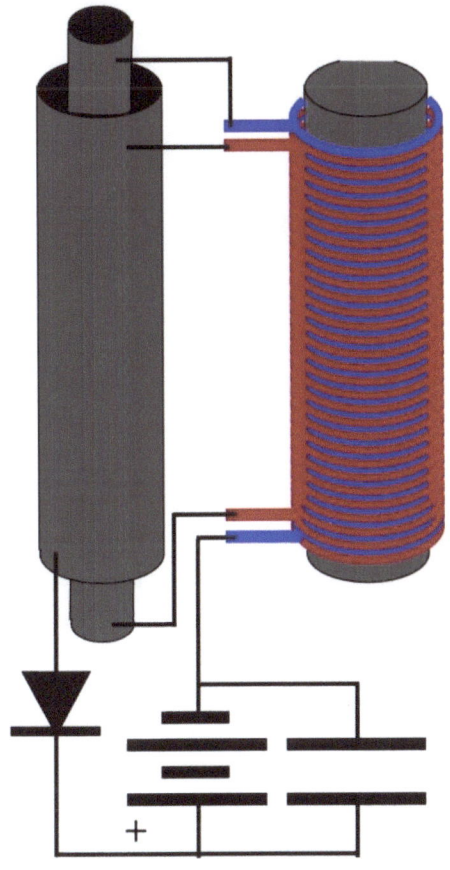

You will find that this creates a high voltage that exceeds what you put into it. Power can be drawn off of your source indefinitely. Infinite power at your fingertips. Go crazy! This recharges your source as power is drawn off of it, so it is an inexhaustible power supply. Imagine the possibilities. Mobile devices and toys with permanent batteries, electric automobiles that run freely, home power plants, planes that never need to land to refuel. It is truly a world of free energy, and it all comes from the cosmic energy, the stars and the center of the galaxy. The ether valve is literally the sought after amplifier that takes in free energy from space and adds it into your circuit. The source battery is never drained and the power it uses is sourced from the ether. One charge will last a lifetime. As Tesla once exclaimed, "See the excitement coming!"

How Does It Work?

The EVe works by receiving the longitudinal magnetic waves that flow from the poles of the Earth from the Sun. It is inexhaustible and will run 24/7 without fail. The battery and the tubes are absolutely necessary. The other components are for technical purposes. The tubes may be constructed of iron or copper. The tubes act as an electrical resonator cavity activated by the magnetic waves. This resonates the battery which is why we get the amplified DC voltage related to the voltage of the battery. The diode prevents the battery from shorting through the coils. The coils themselves are resonated by the circuit. One is energized by the circuit, while the other is used to produce usable power from the first (having more windings in this case). The capacitor stores charge from the circuit to stabilize the output and acts as a tank tuning the circuit. Any tank circuit added to the circuit between the tubes will resonate and create a wireless energy force field. An inductor may be added to further stabilize the output current. There are actually two ways to construct tanks. Here are the circuits consisting of a capacitor and an inductor or coil.

Radiant Energy

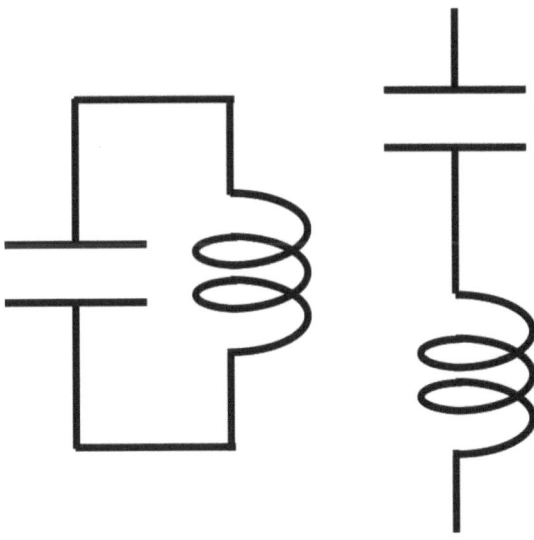

Using your EVe

The actual circuit may vary somewhat from the schematic and still work. The circuit is very forgiving. A voltage divider will suffice to use your EVe. This power is drawn directly from the Ether. Now your voltage divider will not function like a normal voltage divider. Resistances will behave between your battery source and your valve voltage, both need to be considered. Expect odd values somewhere above and below your battery source voltage. The example given below is in parallel to a 6 volt battery with equal coil turns measured from an oscilloscope.

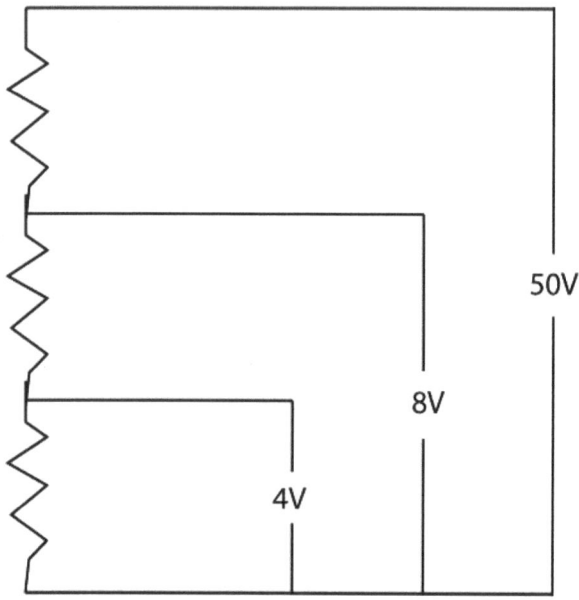

Note that these high voltages will not appear to most DMMs because they are ether currents and not electron currents. The DMM will however register the true voltage of the source battery and indicate that it is indeed slowly charging up over time. The ether currents are completely usable, however, in conventional circuits, and completely safe, offering zero electronic amperage. The only requirement to a functioning circuit is the source battery charge, for conversion purposes. The 50 volts in the given example are real, usable potential and can be discharged into real current through an inductor.

The EVe can be used much like a solar cell, to charge battery banks, coils, or capacitors parallel to the source. In fact, if attached to a solar panel, the panel will begin resonating and emitting usable energy. This way a small battery can be used for

Radiant Energy

high voltage tasks like crossing spark gaps, energizing coils, running large motors, producing alternating current, etc. Proper storage of charge in a capacitor or coil will provide a consistent high voltage.

Attaching a series of diodes or a neon bulb will behave in much the same way as in a Radiant Energy circuit. Realize that the EVe is also a Radiant Energy circuit. This is etheric current mixed with true DC. All circuits attached to the tubes will resonate. Extra windings may be added to the coils for additional outputs. An interesting pulsing ringing effect can be had by placing a transformer across the battery. As with the Joe Cell, water may be added to the chamber to form oxyhydrogen gas for combustion. This makes the cell extra functional. Manipulating the number of turns of the coil will adjust the field effects from the coils.

Looking Toward Applications

Aside from home power plants decentralizing the grid and reducing pressure on outdated utilities, there are other applications of these technologies that may be of great use. Electric vehicles are one that comes to mind. The Joe Cell like applications of the EVe with oxyhydrogen would allow for adaptation of existing combustion engines. Another is power packs for mobile devices. Yet another is sensor technology that can operate independently with minimal maintenance.

The carbon footprint of any carbon intensive infrastructure can easily be eliminated by the use of these technologies scaled up for utility use such as commercial, manufacturing or civil use. This is a great energy solution, and will likely enable intergalactic civilization if scaled up.

Helping out the poor is also an application that comes to mind. Those in need can also benefit from this technology. Free energy is inexpensive and public housing, developments, and public shelters can be outfitted with heating such as low frequency infrared band fields during the winter months, and cooling high frequency band fields during the summer months at a low maintenance cost. This lowers the burden on taxpayers and provides much needed relief to those in need.

Agriculture can benefit from this technology greatly. Artificial light field generation can enable crops to be grown nearly anywhere, even in dark structures or the depths of outer space with minimal energy consumption. In fact, artificial light fields and auroras can be used as an energy storage medium. The power applications will enable running of machinery, fertilization of fields, manipulation of weather, and even the control of pests all via electricity. This will lower the price of organic food and allow for abundance in every sense of the word.

Education and psychology can both benefit from the mental aspects of ether fields of various frequencies. Healthcare can also benefit greatly. This will allow for the direct manipulation of the human mind and body via the aura and chakras, or the human electromagnetic field. Pulsed EMF can have profound psychological and health effects, and a force field of electricity would likely have even greater effects. Hospitals could transmit

Radiant Energy

healing energies to their patients. Enlightenment can be broadcast to raise the level of intelligence of elementary students, or conversely, calmness could be transmitted to mentally unstable individuals in a psychiatric ward. Healing the mind becomes possible because the mind can be directly influenced by the etheric energy on an energetic level. Healing and enlightening populations becomes realistic because energies can be transmitted to the individuals in a totally immersive field over large areas.

Space is the final frontier. Space mining will end all war on this planet. War for resources is what plagues our planet. If resources are mined from space, war becomes irrelevant. Oil is outmoded. Precious metals can be cheaply mined from asteroids. Energy can be obtained in limitless quantities. Special physics like electrogravity (Thomas Townsend Brown effect), force fields, EM drive and tractor beams then become realistic and achievable. Space exploration is no longer a difficulty left to awkward rocketry. Electricity is king!

The Struggle Is Real

Now this won't change things overnight. No matter how viral this book becomes, no matter how many people begin broadcasting free power, the fight for free energy technology will face its enemies. Politics always get in the way of any change in civilization for its betterment. There are vested interests at play that will want to see this technology fail. Expect to face that. Realize that this technology has been personally channeled, built, tested, and thoroughly examined by the author personally and it does work. Everything Gerry Vassilatos claims about Tesla's ether technology is true and tested, and I highly suggest his book *Secrets of Cold War Technology* as a companion manual. However, please use low voltages because we don't want to rule the world, we just want to power our homes and businesses on a grassroots level.

It is however, entirely possible to power a small community on a governmental level. Putting this technology into use in civil infrastructure as a public service would help fulfill Tesla's dream. Bring this up at your local club meetings, share this with your neighbors and family members. You are the resistance! We can light the world one person at a time. One device at a time. One home at a time. Remember, wireless power, true free energy is built on wireless mesh networking of generators and devices. If we share a frequency, we are all connected in unity contributing to the same world field.

Thank you for reading this manual. Your contribution to global change is real and appreciated. Now go out and make this happen. Study and teach this and change can and will happen. Build upon what this manual has offered. Discuss, research, and collaborate. We can bring Tesla's free energy technology to the world if we work together.

Radiant Energy

Works Cited

Vassilatos, Gerry. Secrets of Cold War Technology.

Meyers, Bryant. PEMF – The Fifth Element of Health.

Notes:

www.ingramcontent.com/pod-product-compliance
Lightning Source LLC
Chambersburg PA
CBHW041121180526
45172CB00001B/370